I0484854

SUCCEED IN NUMERACY TESTS

VOLUME 1 – NUMBER CONCEPTS AND THEORY

JASMIN C. ALEXANDER

This book belongs to

_____

School:_____

Class: _____

❖*Practice makes perfect…*

❖*Your attitude determines your altitude…*

TABLE OF CONTENTS

Chapter 2 USING MULTIPLES

Chapter 3 USING FACTORS

Chapter 4 USING PLACE VALUE AND VALUE

Chapter 5 NUMBER SEQUENCE

Chapter 6 COMPARING NUMBERS

Chapter 7 ORDERING NUMBERS

Chapter 8 ROUNDING – OFF NUMBERS

INTRODUCTION

This book is primarily intended to assist students in being successful in mastering Number Concepts up to the Grade 6 level. It covers all of the required areas of Number Concepts in the Primary School Curriculum needed for GSAT Mathematics Exam and serves as the foundation towards helping students to understand Mathematics and its various concepts.

The topics are presented in an easy to understand format with exercises at the end of each of the concepts taught to evaluate students' understanding of the concepts and to help the teacher to be more cognizant of which concepts the students have mastered and which concepts they need more practice.

The chapters are arranged under several headings, each dealing with concepts that are related in one way or the other. Students are thus guided sequentially and logically towards developing a keen understanding of each concept and its relationship to another.

Chapter One deals with understanding numbers and touches on topics such as the Ten Basic Numerals that make up all numbers, the Base Ten System, Counting Numbers, Whole Numbers, Odd and Even Numbers, Ordinal Numbers and Roman Numerals.

Chapter Two deals with 'Using Multiples' and looks at Multiples and concepts that involves the use of multiples, such Lowest Common Multiples, Square Numbers, and Square Roots.

Chapter Three focuses on 'Using Factors' and covers such related concepts as Prime Numbers, Composite Numbers, Common Factors and Highest Common Factors.

Chapter Four looks at Using Place Value and Value and guides students towards using their understanding of these concepts to spell numerals names and write numeral up to billions.

Chapter Five deals with 'Number Sequencing', while **Chapter Six** and **Chapter Seven** deal with 'Comparing Numbers' and 'Ordering Numbers' respectively.

Chapter Eight looks at 'Rounding–Off Numbers'. Students are taught to round –off numbers to the nearest 10s, 100s and 1000s. They are also shown further tips on how to round–off ANY number including rounding-off numbers to their nearest 10th etc., and to the nearest whole number.

The end of the book presents two *'Mastery Tests on Number Concepts'* and *a Certificate of Achievement* which can be presented to each student who has achieved a satisfactory pass in each of these exams.

Alternatively, teachers or parents can select relevant topics from the book for teaching to their students or children in lower grades and then test them on those concepts. Students/children who pass can then be presented with a *'Certificate of Achievement'* at their respective grade levels.

NUMBER CONCEPTS

Chapter 1 UNDERSTANDING NUMBERS

1.1 The 10 Basic Numerals

The 10 basic numbers used in Mathematics are 0, 1, 2, 3, 4, 5, 6, 7, 8 and 9. These numbers were adopted from the *Hindu-Arabic* Number System and are used to show the digit(s) of any number regardless to how big or small it is.

1.2. The Base 10 System

The numbers we use follow the base 10 system. This can be seen as a game where 10 is the key number needed to go from one column to another column (Place Value). A group of ten (10) ones is equal to one (1) ten. Each column's value is thus ten times (10x) as much as the one column below it. These columns' values increase from right to left.

Ten Thousand	Thousand	Hundred	Tens	Ones
				1
			1	0
		1	0	0
	1	0	0	0
1	0	0	0	0

1

$1 \times 10 = 10$

$10 \times 10 = 100$

$100 \times 10 = 1000$

$1000 \times 10 = 10,000$

The biggest number or digit that can be placed in any column is nine (9)

Ten Thousand	Thousand	Hundred	Tens	Ones
				9
			9	9
		9	9	9

1.3 Counting Numbers

Counting Numbers are numbers like 1, 2, 3, 4, 5, 6, 7, 8, 9, 10, 11 etc. Each number is one (1) more than the number before it. Example:

$$1 + 1 = 2 \qquad 2 + 1 = 3 \qquad 3 + 1 = 4 \qquad 4 + 1 = 5$$

Let us Count:

1	2	3	4	5	6	7	8	9	10
11	12	13	14	15	16	17	18	19	20
21	22	23	24	25	26	27	28	29	30
31	32	33	34	35	36	37	38	39	40
41	42	43	44	45	46	47	48	49	50
51	52	53	54	55	56	57	58	59	60
61	62	63	64	65	66	67	68	69	70
71	72	73	74	75	76	77	78	79	80
81	82	83	84	85	86	87	88	89	90
91	92	93	94	95	96	97	98	99	100
101	102	103	104	105	106	107	108	109	110
111	112	113	114	115	116	117	118	119	120

Look carefully at the numbers again. What did you notice about the numbers in each row and in each column?

Fill in the empty boxes with their correct numbers.

If we are counting backward, then the numbers will be one (1) less than the number after it. Count backward from right to left

$\longleftarrow$

1, 2, 3, 4, 5, 6, 7, 8, 9, 10, 11, 12, 13, 14, 15, 16, 17

Exercise 1:

Let us count some bigger numbers. Fill in the missing numbers.

901	902	903		905	906		908	909	
911		913	914	915		917	918		920
921	922		924		926	927	928	929	930
	932	933		935	936		938	939	
941		943	944		946	947		949	950
	952		954	955		957	958	959	
961		963		965	966		968		970
971	972	973	974		976	977		979	
	982		984	985		987	988		990

991		993	994	995	996		998		1000
	1002	1003		1005	1006	1007			1010
1011	1012		1014		1016		1018	1019	1020
									1040

Write the missing counting numbers for each group of numbers below

A. ___, 32, _____   B. _____, 99, _____   C. 198, _____, _____

D. 544, _____, _____   E. _____, _____, 601   F. 300, _____, _____

G. 1,100, _____, 1002, _____   H. 40, ___, ___,   I. _____, _____, 5010

1.4 Whole Numbers

The numbers that we have looked at so far above are all whole numbers. The lowest whole number is zero (0). Numbers that are less than one (1) whole are shown with a decimal point in front of them or are expressed as fractions. For example:

Whole Numbers: 0, 6, 12, 35, 497, 3019, 1123

Numbers less than one whole: 0.35, 0.215, ½, 9/10

Some numbers contains both whole numbers and non-whole numbers (Fractions or decimals). Some examples are:

12.40
wholes   decimal

6 ¼
wholes   fractional part

The decimal point separates the whole number from the part of a whole (fraction).

Exercise 2:

Circle the whole numbers from the list of numbers below:

124	1890	0.5	2/4	24	7
96	563	7/10	0.172	5403	5

1.5 Odd Numbers and Even Numbers

1	2	3	4	5	6	7	8	9	10
odd	even	odd	even	odd	even	odd	even	odd	even

- Can you see a pattern here? _____

- What do you noticed about the

  numbers?_____

  _____

1.5.1 Odd Numbers – An Odd Number is a number that cannot go into groups of two (2) without a remainder. (e.g. 1, 3, 5, 7, 9, 41, 673)

Any number that ends with a digit that is an '**Odd Number**' (i.e. 1, 3, 5, 7 or 9) is itself an odd number. For Example:

$$2\underline{1}, \quad 34\underline{5}, \quad 782\underline{9}, \quad 12\underline{3}, \quad \text{and} \quad 4,56\underline{7}$$

These numbers all end with a digit that is an Odd Number; therefore, they are all 'Odd Numbers.

1.5.2 Even Numbers – An Even Number is a number that can go into groups of two (2) without a remainder. (E.g. 2, 4, 6, 8, 10 –*Think of counting in 2s*)

Any number that ends with a digit that is a digit that is an '**Even Number**' (e.g. 2, 4, 6, 8) or the digit '0' (if it is a multiple of ten) is itself an Even Number. For Example: 3$\underline{2}$, 34$\underline{6}$,   782$\underline{4}$,   12$\underline{8}$,   2,50$\underline{0}$ and   4,56$\underline{0}$

These numbers all end with an Even Number or the digit '0' showing it is a multiple of ten; therefore, they are all 'Even Numbers.

Exercise 3:

Write **Odd** or **Even** to describe each number below:

a. 13 _____   b. 579 _____   c. 948 _____

d. 1,002 _____   e. 50 _____   f. 2345_____

g. 6 _____   h. 707 _____   i. 122 _____

1.6. Ordinal Numbers – These numbers show positions. Each ordinal number has a number and the last two letters in its name.

1st	first	11th	eleventh	21st	twenty-first
2nd	second	12th	twelfth	30th	thirtieth
3rd	third	13th	thirteenth	40th	fortieth
4th	fourth	14th	fourteenth	50th	fiftieth
5th	fifth	15th	fifteenth	55th	fifty-fifth
6th	sixth	16th	sixteenth	100th	one hundredth
7th	seventh	17th	seventeenth		
8th	eighth	18th	eighteenth		
9th	ninth	19th	nineteenth		
10th	tenth	20th	twentieth		

Exercise 4:

Look carefully at the spellings of the ordinal numbers; then answer these questions:

1. What do you noticed about most of their endings?_____

2. What is the difference in spellings of the numbers:

a. 9 and 9th

_____

b. 12 and 12th

_____

c. 20 and 20th

_____

3. Myra ran a race and came 24th. What did the girl in front her came?_____

  4. Susan won the race. What did she come in the race?

_____

1.7 Roman Numerals – These are numbers represented by certain letters. These letters may be written in *capital* or *common letters*. The 'key letters' used are:

I = 1   V = 5   X = 10   L = 50

C =100   D = 500   M = 1000

The rules of this numbering method are:

➢ Only a maximum of three (3) similar letters can be used consecutively to represent a number or part of a number.

➢ When a number needs more than three (3) similar letters coming one after the other in a row, you must go to the next key letter of a higher value and subtract the difference to arrive at the number you wanted. For example, 3 equal iii but 4 **does not** equal iiii; therefore you have to go to the next key number 'v' which is equal to 5 and take one from it to get 4.

➢ To take away a number (i.e. one or i), you put it in the front of the number from which you are going to take away. Therefore, 4 is equal to 5 take away 1; that is 4 = iv

➢ To add a number, you put it after the number to which it is being added. Therefore, 6 is equal to 5 + 1; that is 6 = vi

Study the **Roman Numerals** below and the **Hindu-Arabic Numbers** (i.e. the numbers we usually use to write numbers) that they represent:

HINDU-ARABIC NUMERALS	ROMAN NUMERALS	WORKINGS
1	I	1
2	II	1+1
3	III	1+1+1
4	IV	5 -1
5	V	5
6	VI	5 +1
7	VII	5 + 2
8	VIII	5 + 3
9	IX	10 -1
10	X	10

Exercise 5A: Copy and complete the tables below correctly?

HINDU-ARABIC NUMERALS	ROMAN NUMERALS	WORKINGS
1		
2		
3		
4		
5		
6		
7		
8		
9		
10		
11	XI	10 + 1
12	XII	10 + 2
13	XIII	10 + 3
14	XIV	10 + 4
15	XV	10 +5
16	XVI	10 + 6
17	XVII	10 + 7
18	XVIII	10 + 8
19	X IX	10 + 9
20	XX	10 + 10
21	XXI	10+ 10 +1

HINDU-ARABIC NUMERALS	ROMAN NUMERALS	WORKINGS
22	XXII	10 +10 + 2
23		10 +10 + 3
24		10 +10 + 4
25		10 +10 + 5
26		10 +10 + 6
27		10 +10 + 7
28		10 +10 + 8
29		10 +10 + 9
30		10 +10 + 10
31		10 +10 + 10 + 1
32		10 +10 + 10 + 2
33		10 +10 + 10 + 3
34		10 +10 + 10 + 4
35		10 +10 + 10 + 5
36		10 +10 + 10 + 6
37		10 +10 + 10 + 7
38		10 +10 + 10 + 8
39		10 +10 + 10 + 9
40		50 − 10
41		(50 − 10) + 1
42		(50 − 10) + 2

Exercise 5B:

Compare these Roman Numerals using the < (less than) Sign and the > (more than) Sign

a). xi ☐ ix   (b). vii ☐ viii

(c). xix ☐ xxi   (d). XL☐ LIX

1.8 Writing Larger Roman Numerals

> The letter 'M' represents 1000 in Roman Numerals. It is the largest 'key letter.' Roman Numerals greater than 3000 are therefore written in their regular way with a line above to represent thousands. For example:

$$4000 = \overline{iv} \qquad 5000 = \overline{v} \qquad 6000 = \overline{vi}$$

Exercise 5C:

Write the Roman Numeral for these Hindu-Arabic Numerals below:

HINDU-ARABIC NUMERALS	ROMAN NUMERALS	WORKINGS
100		
200		
300		
400		
500		
600		
700		
800		
900		

1,000		
1,200		
2,020		
2,240		
4,000		
10,000		

Chapter 2 USING MULTIPLES

2.1 Multiplication Table: Multiplication is repeated addition of the same number. Therefore in:

➢ one times (1x) tables, 1 is added each time to the number before it

➢ two times (2x) tables, 2 is added each time to the number before it

➢ three times(3x) tables, 3 is added each time to the number before it and so forth.

Study the multiplication table below

1x	2x	3x	4x	5x	6x	7x	8x	9x	10x	11x	12x	13x
1	2	3	4	5	6	7	8	9	10	11	12	13
2	4	6	8	10	12	14	16	18	20	22	24	26
3	6	9	12	15	18	21	24	27	30	33	36	39
4	8	12	16	20	24	28	32	36	40	44	48	52
5	10	15	20	25	30	35	40	45	50	55	60	65
6	12	18	24	30	36	42	48	54	60	66	72	78
7	14	21	28	35	42	49	56	63	70	77	84	91
8	16	24	32	40	48	56	64	72	80	88	96	104
9	18	27	36	45	54	63	72	81	90	99	108	117
10	20	30	40	50	60	70	80	90	100	110	120	130
11	22	33	44	55	66	77	88	99	110	121	132	143
12	24	36	48	60	72	84	96	108	120	132	144	156
13	26	39	52	65	78	91	104	117	130	143	156	169
14	28	42	56	70	84	98	112	126	140	154	168	182
15	30	45	60	75	90	105	120	135	150	165	180	195

Try to learn and memorize at least your 1x to 12x tables. It will help you later on the solving many mathematical problems quickly and it will also exercise your brain cells.

2.2 Multiples – Multiples are repeated addition of the same number (think skip counting by the same number or multiplication tables). For example, five (5) multiples of two are:

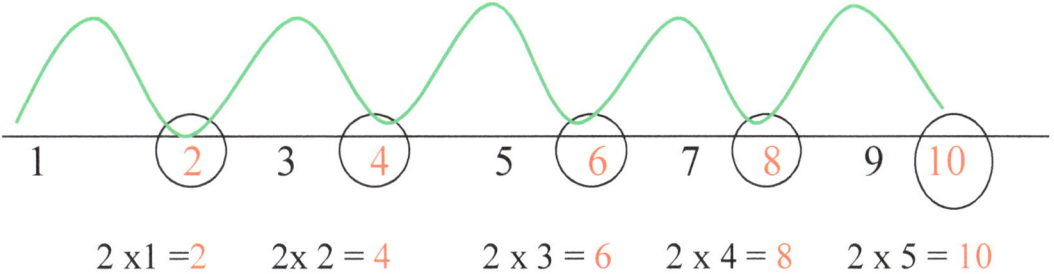

2 x1 =2   2x 2 = 4   2 x 3 = 6   2 x 4 = 8   2 x 5 = 10

Five multiples of 2 = 2, 4, 6, 8, 10

Five multiples of 3 = 3, 6, 9, 12, 14

Five multiples of 4 = 4, 8, 12, 16, 20

Exercise 6:

Write three multiples of each number below:

5 = ____,____, ____   9 = ____,____, ____   10 = ____,____,____

6 = ____,____,____   17 = ____, ____, ____   8 = ____, ____, ____

11 = ____,____,____   12 = ____,____,____   13 = ____, ____, ____

2.3 Lowest Common Multiple (L.C.M.) – The L.C.M. is the lowest multiple that is the *same* for two (2) or more give numbers. Example:

Method 1

What is the L.C.M of 2 and 3?

Multiples of 2 = 2, 4 ⬭6⬭   Multiples of 3 = 3 ,⬭6⬭, 9

Therefore, **6** is the L.C.M. of 2 and 3 because it is the lowest multiple that is the <u>same</u> in both of them.

Example: *What is the L.C.M. 2, 3 and 4?*

Multiples of:

2={2,4,6,8,10,**12**}   3= {3, 6, 9,**12**,15}   4= {4, 8, **12,** 16}

Answer: The L.C.M. of 2, 3 and 4 is 12

Method 2:

➢ The L.C.M of a group of numbers may also be found by looking consecutively at **multiples** of the bigger/biggest number of the group, and then multiplying each of the other numbers in the group by a number to arrive at that **multiple**. When each of the other number arrives at the same multiple, that multiple is the L.C.M. of the group of numbers.

Using the same question above:

What is the L.C.M. 2, 3 and 4?

The largest number of the group is 4.

❖ Some multiples of 4 = 4, 8, 12, 16

Other numbers of the group are 2 and 3

❖ Some multiples of 2 are: 2, 4, 6, 8, 10, 12, 14…

❖ 3 cannot be multiplied by any number to get 4 or 8, so, neither 4 nor 8 is the L.C.M, of the group.

❖ Some multiples of 3 are: 3, 6, 9, 12, 15…

All of the numbers in the group shows 12 as the *lowest* **_same_** *multiple*. Therefore, 12 is the L.C.M. of 2, 3 and 4.

Method 3:

➤ The L.C.M. can also be found by using the '***Prime Factorization***' method. To do this, a table is drawn to include the numbers that you have to find the L.C.M. of, and these numbers are divided consecutively by prime numbers by which they are divisible until they have all reached their minimum of 1. The prime numbers used are then multiplied to find the L.C. M. For example:

What is the L.C.M. of 2, 6 and 8?

÷

2	2	6	8
2	1	3	4
2	1	3	2
3	1	3	1
	1	1	1

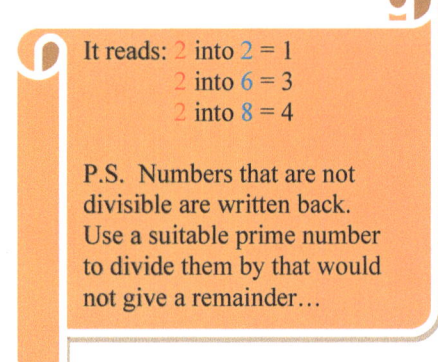

It reads: 2 into 2 = 1
2 into 6 = 3
2 into 8 = 4

P.S. Numbers that are not divisible are written back. Use a suitable prime number to divide them by that would not give a remainder…

L.C.M. = 2 x 2 x 2 x 3 = 24

*Note: The L.C.M of any group of number will **never be smaller** than the bigger/biggest number of that group.*

Exercise 7

Find the L.C.M. of the numbers below

a) 2 and 8   b) 3, 4 and 12   c) 4 and 5

d) 6 and 7   e) 5, 6 and 15   f) 9 and 18

g) 62 and 34   h) 12, 26 and 38   i) 125 and 340

2.4 Square Numbers: A square number is the answer arrived at when a number is multiplied by itself. For example:

3 x 3 = 9, therefore 9 is a square number.

2 x 2 = 4, therefore 4 is a square number

4 x 4 = 16, therefore 16 is a square number

When a square number is shown by dots or similar objects, the arrangement forms a squared pattern. For example:

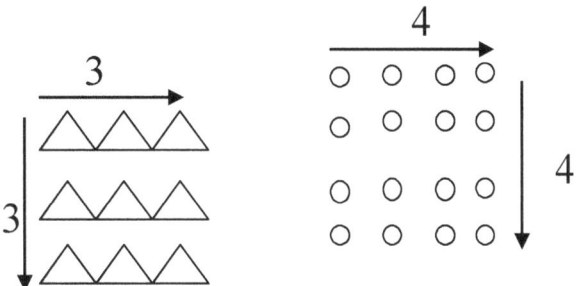

2.5 Square Root: The opposite of the square of a number, is its square root.

➢ Square root is represented by the symbol $\sqrt{}$

➢ To find the square root of a number, ask yourself: *What number if multiplied by itself will give you the number you want to find the square root of?*

For example:

$$\sqrt{16} = 4 \quad \text{(Remember 4 x 4 = 16)}$$

$$\sqrt{25} = 5 \quad \text{(Remember 5 x 5 = 25)}$$

$$\sqrt{36} = 6 \quad \text{(Remember 6 x 6 = 36)}$$

Exercise 8

Complete the table below correctly:

Number	Square number	Square root $\sqrt{}$
2 x 2	4	2
	9	3
4 x 4	16	
5 x 5		5
6 x 6	36	
7 x 7	49	
8 x 8		8
9 x 9		
10 x 10	100	
12 x 12		12
13 x 13		
14 x 14		

2.6 Exponential Form – The exponential form of a number has two parts; namely a **base** (the bottom number) and a **power** or **exponent** (the small number above the right of the base).

❖ The base shows the factor that is being repeated.

❖ The power or exponent tells the number of times the factor is repeated.

For example: 5 x 5 written in its exponential form is:

exponent or power   *(the number of times the base is repeated)*

$$5^2$$

base   *(the factor that is repeated)*

Some Rules about Exponential Form Numbers:

❖ Any number raised to the 0 power is equal to 1 (e.g. $2^0 = 1$, $16^0 = 1$)

❖ Any number raised the power '1' or the first power is equal to the same number (e.g. $6^1 = 6$, $42^1 = 42$ etc.)

❖ Numbers with the power '2' are said to be raised to the second power or squared (e.g. 4^2 is read as 4 to the second power or 4 squared)

❖ Numbers with the power '3' are said to be raised to the third power or cubed (e.g. 5^3 is read as 5 to the third power or 5 cubed)

❖ Numbers with the power '4' are said to be raised to the fourth power.

❖ Numbers with the power '5' are said to be raised to the fifth power.

Exercise 9

Complete the table below correctly:

Exponential Form	Factor Form	Product
4^0		
6^1		
7^2	7 x 7	
2^2	2 x 2 x 2	
3^5		
10^7		
2^6		

Chapter 3 USING FACTORS

3.1 Factors – The factors of a number are all the numbers that can be divided into it without leaving a remainder. Factors can also be seen as numbers which are multiplied to give a product. Numbers that give the same product when they are multiplied are <u>factors</u> of that product. Example:

$$1 \times 4 = 4 \qquad\qquad 2 \times 2 = 4$$

factor   factor   product   factor   factor   product

The factors of 4 are: 1, 2, and 4

Exercise 10:

List all the factors of each numbers below:

i. 2 = {_____}

ii. 7 = {_____}

iii. 9 = {_____}

iv. 15 = {_____}

v. 16 = {_____}

vi. 20 = {_____}

vii. 18 = {_____}

viii. 36 = {_____}

3.2 Prime Number

3.2 Prime Number – A Prime Number is a number that has only two factors; itself and 1. For example: 2 is a prime number because only 2 x 1 or 1 x 2 can be multiply to give 2 . That is, only the number 2 and 1 give the product 2.

Some other examples of Prime Numbers are 3, 5, 7, 11, 13, 17, 19, 23, 29, 31…

3.3 Composite Numbers

3.3 Composite Numbers – A composite number is any number that has more than two (2) factors. For example: 4, 6, 8, 9, 10, 12, 14, 15, 16, 18 etc.

➤ What are the factors of 4?   (That is, what are the different numbers that can be multiple to give the product 4?)

1 x 4 = 4
2 x 2 = 4   Factors of 4 are {1, 4 and 2}

➢ What are the factors of 12?

1 x 12 = 12
2 x 6 = 12
3 x 4 = 12   Factors of 12 are {1, 12, 2, 6, 3 and 4}

3.4 Highest Common Factor (H.C.F.): The H.C.F. is the biggest or greatest factor that is the same for a set of numbers.

Using the Common Factor Method:

To find the H.C.F. of a given set of numbers:

Step 1:
❖ Find the **factors** of each of the given numbers;

Step 2:
❖ Identify all the **common factors** (i.e. factors that are the same) for each of the given numbers;

Step 3:
❖ Choose the **biggest common factor** as your answer. For example:

Question: What is the H.C.F. of 15, 60 and 90?

Step 1: Find the **factors** of each of the given numbers:

➢ Factors of 15 = {1, 15, 3, 5}

➢ Factors of 60 = {1, 60, 2, 30, 3, 20, 4, 15, 5, 12, 6,10}

➢ Factors of 90 = {1, 90, 2, 45, 3, 30, 5, 18, 6, 15, 9, 10}

Step 2: Identify all the **common factors** (that is, factors that are the same) for each of the given numbers:

➢ The common factors of 15, 60 and 90 are 1, 3, 5 and 15

Step 3: Choose the **biggest common factor** as your answer:

➢ The number 15 is the biggest common factor amongst the three given numbers. Therefore, 15 is the Highest Common Factor of the numbers 15, 60 and 90.

Using the Prime Factorization Method:

Another method that may be used to find the H.C.F. is the *'Prime Factorization Method'*. When using this method to find the H.C.F. of given numbers, the prime factors of each given number are found separately and then the similar occurrences of prime number patterns (i.e. index or powers) are multiplied to arrive at the H.C.F.

For example: **What is the H.C.F. of 12 and 24?**

÷

2	12
2	6
3	3
	1

÷

2	24
2	12
2	6
3	3
	1

12 = 2 x 2 x 3   **24** = 2 x 2 x 2 x 3

Similar occurrences of prime number patterns are **2 x 2 x 3,** therefore the H.C.F. = 2 x 2 x 3 = 12

Exercise 11

Write the Common Factors (C.F.) for each group of numbers below:

i) 3, 6 and 12   (ii) 4, 12 and 16

iii) 7, 21 and 28   (iv) 10 and 25

Exercise 12

Find the Highest Common Factor (H.C.F.) of each set of numbers below:

i) 12 and 34   (ii) 8, 14 and 20

iii) 9, 36 and 6   (iv) 32, 16 and 8

(v) 16, 18 and 27   (vi) 5, 6 and 7

Chapter 4 USING PLACE VALUE AND VALUE

4.1 Place Value: This tells **where** the digit is placed. Some examples of place value are Ones, Tens, Hundreds, Thousands etc. The place value table below shows the place value of the number 1234.56:

Thousand	Hundred	Tens	Ones	Tenths	hundredths	Thousandths
1	2	3	4	• 5	6	3

The place value of the underlined digit:

<u>1</u>234.56= Thousands   1<u>2</u>34.5= Hundreds

12<u>3</u>4.56 Tens   123<u>4</u>.56 = Ones

1234.<u>5</u>6= tenths   234.5<u>6</u> = hundredths

NB. The decimal point separates the wholes from the fraction (part of a whole).

Exercise 13:

A. State the **Place Value** of the underlined digits below:

a) 18<u>0</u> =   b) 7<u>9</u>33 =   c) 5444.9<u>8</u> =

d) <u>3</u>000.12 =   e) 22.<u>5</u> =   f) 5<u>4</u>7.75 =

In the number **4653.92**, what is the place value of the digit:

a) 2   b) 3   c) 4
d) 5   e) 9   f) 6

a) 2437.8   b) 3257.86   c) 469.57   d) 94.73

The table below shows examples of place value for whole numbers from **ones** to **millions**

Millions	Hundred Thousands	Ten Thousands	Thousands	Hundreds	Tens	Ones
					1	0
				1	0	0
			1	0	0	0
		1	0	0	0	0
	1	0	0	0	0	0
1	0	0	0	0	0	0

Remember: The place value of each digit is determined by its position.

4.2 Value: This tells how much the digit is worth.
Example

Thousand	Hundred	Tens	Ones	Tenths	Hundredths
1	2	3	4	• 5	6

The value of the underlined digit:

1234.56= 1000   (1 x 1000)

1234.5 = 200   (2 x 100)

1234.56 = 30   (3 x 10)

1234.56= 5/10 or 0.5   (5 x 1/10)

Exercise 14 A:

A. State the **Value** of the underlined digits below:

a) 180 =   b) 7933 =   c) 5444.98 =

d) 3000.12 =   e) 22.5 =   f) 547.75 =

g) 45, 309, 087 =   h) 67.34567 =   i) 608,430 =

4.3 Expanded Notation: Writing a number in its expanded form is the same as writing the value for each of the standard number's digit. For example:

What is the expanded notation of the number 4530?

Th. H. T. U
4 5 30 = 4000 + 500+ 30 + 0

Exercise 14 B

Write these numbers in their expanded form:

    i)   267 =

    ii)   9031 =

    iii)   77245 =

    iv)   3691005 =

Exercise 14 C

Write these numbers in their standard form:

    i)   200 + 10+ 4 = _____

    ii)   6000 + 300 + 1 = _____

    iii)   90 + 900 + 3000 + 4 = _____

    iv) $(5 \times 1000) + (8 \times 100) + (5 \times 10) =$ _____

4.4 Spelling Number Names

Learn to spell these numeral names below correctly. Then let your teacher or a friend test you. (N.B. *The number names are broken into syllables to help you to remember how they are spelt more easily*)

1 – one	10 – ten	19 – nine teen	100 – one hun dred
2 – two	11 – e le ven	20 – twen ty	1000 – one thou sand
3 – three	12 – twelve	30 – thir ty	1,000,000 – one mil li on
4 – four	13 – thir teen	40 – for ty	1,000,000,000 – one bil li on
5 – five	14 – four teen	50 – fif ty	
6 – six	15 – fif teen	60 – six ty	
7 – se ven	16 – six teen	70 – se ven ty	
8 – eight	17 – se ven teen	80 – eigh ty	
9 – nine	18 – eigh teen	90 – nine ty	

➢ Numbers 1 to 9 are one digit numbers.

➢ Numbers from 10 to 99 are called two digits numbers because they have two digits. For example, the number 10 has two digits. They are 1 and 0. The digits in these numbers are spelt as the number they represent and not separately by their digits. For example:

13 = thirteen   (not one three)

68 = sixty-eight (not six eight)

➢ Numbers from 100 to 999 are called three-digit numbers because they each contain three digits. For example a hundred has a one and two zeros (1 0 0). When you are spelling numbers with three digits, you should spell the value of the digit in the hundreds' column. If there are any other digits besides zero in the Ones and Tens columns, then the word **'and'** is used before spelling those two last digits as a number. For example:

200 = two hundred   206 = two hundred and six
700 = seven hundred   879 = eight hundred and
                                 seventy-nine

➢ Numbers with more than three digits are usually separated by a comma or space. The Comma or space is placed after every three digits from right to left (That is, beginning from the Ones Column). Three examples are: 1, 400, 38, 590 and 14 222.

➢ Digits in the ***thousands, ten thousands*** and ***hundred thousand*** columns can be called the **'Thousands Family**.' Any digit(s) in these columns are spelt collectively as a number then the word thousand is written after so that the digits represents the amount of thousands. For example:

1, 000 = one thousand

23,000 = twenty-three thousand

40, 300 = forty thousand, three hundred

10, 001 = ten thousand and one

When spelling numbers, spell the digit(s) in each 'family' as one number, then add their family names; that is **thousand, million** or **billion**. However, the **Ones** family name is not written at the end of its number name. For example

235 = two hundred and thirty-five

↑
Ones' family

23, 167 = twenty-three **thousand,** one hundred and sixty-seven

↑ ↑
Thousands' *Ones'*
family *family*

45,709,000 = forty-five **millions**, seven hundred and nine
thousand

↑ ↑ ↑
Millions' *Thousands'* *Ones'*
family *family* *family*

Remember: Each family can only have a maximum of three (3) digits.

Number families are grouped from the smallest family (that is, from the 'Ones Family') to the largest family in that number.

Exercise 15:

A. Group these numbers' digits in 'families':

a) 256194 -   _____

b) 183030586 - _____

c) 6895420156 - _____

B. Spell these number names correctly. Use the place value chart and number families to help you

Millions Family			Thousands Family			Ones Family			NUMBER NAMES
HUN MIL	TEN MIL	MIL	HUN. THOU	TEN THOU	THOU.	HUN	TENS	ONES	
						1	0	1	
				9	0	0	4	7	
					4	7	8	3	

			4	2	9	0	0	0	
	5	0	1	1	1	1	1		
4	2	2	3	5	9	9	9		
9	8	1	1	5	6	0	0	0	

4.5 Writing Numerals – Writing numerals or numbers are usually required when the number names are given. For large numbers, you may use a **place value table** or look for the numeral **"Family Names"** to help you to write the numerals correctly.

Exercise 16

Copy and complete the table below correctly

Number Names	MILLIONS FAMILY			THOUSANDS FAMILY			ONES FAMILY			Numerals
	H M	T M	M	H T	T T	T H	H	T	U	
Twenty eight								2	8	28
Five hundred and seventy-four							5	7	4	574
Three thousand nine hundred and one						3	9	0	1	3,901
Ninety thousand and forty					9	0	0	4	0	90,040
Thirty-five thousand six hundred					3	5	6	0	0	35,600

Forty-one thousand and eight									
Seventeen thousand two hundred and fifteen									
Five									
Ninety-six									
Four thousand four hundred and forty									
One million six hundred									
Thirty-five million one hundred									
Twenty-eight million five thousand and four									

Exercise 17:

Write the numeral for each numeral name below

❖ Thirteen = _____

❖ thirty-four = _____

❖ seven hundred and twenty-seven = _____

❖ sixty-two thousand = _____

❖ eighty-one thousand, three hundred and two = _____

❖ five hundred and six thousand , eight hundred and fourteen =_____

❖ eighty-seven million seventy three thousand =_____

❖ forty million forty thousand and forty = _____

❖ sixteen billion four hundred = _____

❖ three thousand and ten = _____

<mark>Try writing these numerals. The first one is done for you.</mark>

❖ nine and two-tenths = 9 2/10   or 9.2   (nine wholes and two-tenths)

❖ five and three-tenths = _____

❖ thirty-seven and sixteen hundredths = _____

❖ one hundred and two and seven thousandths = _____

Chapter 5 NUMBER SEQUENCE

5.1 Sequencing Number: This is a method used to arrange numbers to follow a set operation. To find the number sequence or number pattern you have to look at the numbers given and try to find out what is being done. Ask yourself this question: *Are the numbers increasing or decreasing?*

If the numbers are increasing, as you read them from left to right, then the operation being used is either **Addition** or **Multiplication**.

➢ Look for two numbers that come one after the other. Check to see how many numbers have been added to the smaller number to arrive at the bigger number. Then continue that number pattern to solve the missing number(s). For example:

+2 +2 +2 +2   +2
2, 4, 6, 8, 10, 12   (*N.B. In this number sequence, 2 is being added each time*)

12, 16, 20, 24, 28, 32   (*N.B. In this number sequence, 4 is being added each time*)

2x2 4x2 8x2 16x2
2, 4, 8, 16, 32   (*N.B. In this number sequence, each number is being multiply by 2*)

➢ If numbers are missing at the beginning, these numbers will be smaller. So if the pattern is add 3 each time (from left to right), then it will be the opposite; that is, subtract 3 each time from right to left. For example: 6, 9, 12, 15, 18, 21

➢ If the numbers are decreasing, as you read them from left to right, then the operation being used is either **Subtraction** or **Division**.

➢ Look for two numbers that come one after the other. Check to see how many numbers have been subtracted from the bigger number to arrive at the smaller number. Then continue that number pattern to solve the missing number(s). For example:

-5   -5   -5   -5
25, 20, 15, 10, 5

➢ Sometimes letters are added with the numbers to make the problem more interesting. Follow the same steps above accordingly to find the number pattern. Then use your knowledge of the alphabet to follow the letter pattern attached with each number. For example:

0A, 3B, 6C, 9D, 12E

Can you state what the next two missing numbers for the number series below?

+1 +2 +3 +4 +5
2, 3, 5, 8, 12, _____, _____

Exercise 18:

Complete the number sequences below correctly:

a) 5, 10, ____, ____, ____

b) 3, 9, _____, _____, ____

c) ____. ____. 18, 21, _____

d) 10, _____, 30, _____, 50

e) 2A, 4C, ____, _____, ____

f) 108, 109, ____, _____, ____

g) I, II, _____, _____, _____

h) 9th, 12th, _____, _____, _____

Chapter 6 COMPARING NUMBERS

6.1 Comparing Numbers: Numbers are usually compared by their sizes. The three (3) main symbols used in comparing numbers are:

❖ < which means is **'less than'** *(bend your left hand to make the 'less than' sign)*

❖ > which means is **'greater than'** *(bend your right hand to make the 'more than' sign);* **and**

❖ = which means is ' **equal to**' or the '**same as**'
When comparing numbers start your comparison from the digit with the highest value and work your way downward. Remember to always compare numbers in similar columns with each other. If one number does not have a digit in a similar column to compare, assume the digit to be a zero (0). For example

Use these symbols to compare the numbers below:

<, > or =

Th. H. T. O   Th. H. T. O

2 354 ☐ 2 345   - In the thousands column, they are the same

2 3 54 ☐ 2 3 45   - In the hundreds column they are the same

2 35 4 ☐ 2 34 5   - In the tens column they are different; 5 is **more than** 4

Therefore, the correct symbol is the more than (>) symbol. Your answer reads: 2354 > 2345 (two thousand three hundred and fifty-four is *more than* two thousand three hundred and forty-five)

Exercise 19:

Describe the numbers in each set below by using the symbols <, = or >

a) 256 ☐ 265

b) 210 ☐ 102

c) 46 509 ☐ 46541

d) 10000 ☐ 100000

e) 123 ☐ 406

f) 7893 ☐ 789

g) 11226 ☐ 62211

h) 34.34 ☐ 28.342

i) 8805 – 436 ☐ 9268 – 899

j) 572 + 497 ☐ 615 + 549

Chapter 7 ORDERING NUMBERS

7.1 Ordering Numbers: Numbers can be ordered or arranged according to their sizes. Some numbers are arranged in *ascending order*, which is from *smallest to biggest* while some are arranged in *descending order*, which is from *biggest to smallest.*

❖ *Ascend means to 'go up' -* So start listing numbers from the bottom (lowest) to the top (highest).

❖ **Descend** *means to 'go down'-* So start listing numbers from the top (highest) to the bottom (lowest).

Exercise 20

 Arrange these numbers in ascending and descending order:

i) 809, 372, 756, 1031 and 469:

Ascending order:

_____

Descending order:

_____

(ii) 7502, 7520, 5702, 7700 and 6045

Ascending order:

_____

Descending order:

_____

(iii) 98765, 87643, 111000, 050012, 300211 and 987009

Ascending order:

_____

Descending order:

_____

Chapter 8 ROUNDING-OFF NUMBERS

8.1 Rounding-off Numbers: To round-off a number is like choosing an estimated number that is closest to the value you desire; such as the nearest whole number, tenths, tens, hundreds or thousands etc. Your knowledge of Place Value can help you here.

8.2 Rounding-off Numbers to the Nearest 10s: When rounding off numbers to the nearest tens, any answer chosen must be a multiple of ten (E.g. 10, 20, 30, 40, 50…) or zero (0) if the number is below five.

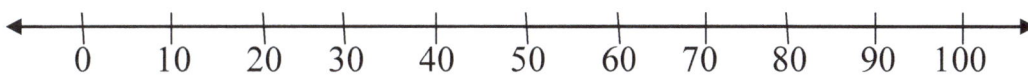

If the number given is not a multiple of ten, then the number's digit in the *Ones column* will either be rounded up if it is half of ten (i.e. 5) or more, or rounded down if it is four or less.

➤ Using the number line as a guide, you will notice that if a given number is not a multiple of ten (10), it will fall between zero and ten or two multiples of ten. For example, the number 12 will fall between 10 and 20 because it is greater than 10 but less than 20.

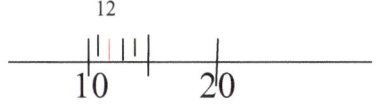

➢ The number 12 is nearer to 10 that it is to 20, so it is rounded down. Therefore, when rounding–off the number 12 to the nearest tens, the answer is 10. (N.B in the number 12, the digit 2 in the ones line is less than half of 10, so the number is rounded down.)

Tens   Ones

$$1 \quad 2 = 10 + \underline{2} \rightarrow$$ The number is said to be rounded down because its value has been reduced. It has been round off to the nearest ten, thus the 10 has been kept and the 2 has been dropped because it is less than half of 10.

➢ When a number is rounded up to the nearest ten, it means that the digit in the ones line has been estimated (rounded off) to a value of ten. This is only done when the digit in the ones line is 5 or more than 5. For example, 46 rounded off to the nearest ten is 50 (N.B. the number 46 is between 40 and 50. It is nearer to 50 than 40; so 46 to the nearest ten is 50)

Tens   Ones

$$4 \quad 6 = 40 + \underline{6} \rightarrow$$ The number is said to be rounded up because its value has increased. The digit in the Ones line is more than four, so it has been estimated (rounded off) to a10.

$$= 40 + \underline{10} = 50$$ Thus, forty plus ten equals fifty

➢ If a number that is given to be rounded–off to the nearest ten is already a multiple of ten (i.e. the digit in the Ones line is '0'), then it value remains the same. For example, the number 60 to the nearest ten is 60.

Exercise 21

Complete the table below:

Numbers	Location Between	Round up/Round down (hint: look at the Ones line digit)	Answers (rounded-off to the nearest 10)
2<u>4</u>	Down Up 20 – 30	Round down	20
7<u>7</u>	70 – 80	Round up	80
123<u>6</u>	1230 – 1240	Round up	1240
88<u>1</u>	880 – 890	Round down	880
<u>8</u>			
9<u>0</u>			
10<u>2</u>			
42, 45<u>9</u>			
620<u>0</u>			
26<u>3</u>			

8.3 Rounding off Numbers to the Nearest 100:

When rounding off numbers to the nearest hundred , any answer chosen must be a multiple of a hundred, or zero if the number is nearer to zero (0) than to one hundred (100)

➤ Numbers that is less than half of a hundred or less than 50 (i.e. their tens line digit is less than 5) would be rounded down. Therefore, numbers 1 to 49 rounded off to the nearest hundred is zero (0) because they are all nearer to zero than a hundred.

➤ Similarly, numbers which have digits in the Tens and One lines being less than fifty (50), will have these digits rounded down to zeros (00). For example:

123 rounded off to the nearest 100 is 100

34549 rounded off to the nearest 100 is 34500

➤ Numbers with digit in the <u>Tens line</u> that is five (i.e. 50) or more are rounded up. These numbers are nearer to the multiple of hundred on their right, so they are rounded off to that number. For example, 393 to the nearest hundred is 400; this is so because the number 393 is found between 300 and 400. It is nearer to the number 400 (on its right); therefore is rounded off to 400.

Exercise 22

Complete the table below:

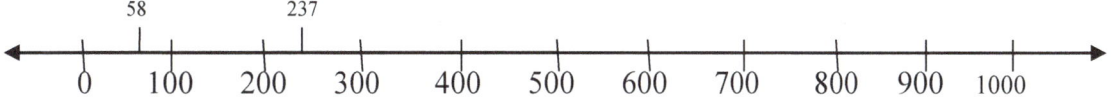

Numbers	Location Between	Round up/Round down (hint: look at the Tens line digit)	Answers (rounded-off to the nearest 100)
	Down Up		
58	0 – 100	Round up	100
237	200 – 300	Round down	200
789	700 – 800	Round up	
999			
1145			
33			
571			

8.4 Rounding off Numbers to the Nearest 1000:

When rounding off numbers to the nearest thousand , any answer chosen must be a multiple of a thousand or zero if the number is nearer to zero (0) than one thousand (1000)

➤ Numbers that is less than half of a thousand or less than 500 (i.e. their Hundreds line digit is less than 5) would be rounded down. Therefore, numbers 1 to 499 rounded off to the nearest hundred is zero (0) because they are all nearer to zero than to a thousand.

➤ Numbers with digit in the <u>Hundreds line</u> that is five (i.e. 500) or more are rounded up. These numbers are nearer to the multiple of a thousand on their right, so they are rounded off to that number. For example:4,567 to the nearest thousand is 5,000; this is so because the number 4,567 is found between 4,000 and 5,000, but it is nearer to the number 5000 (on its right); than it is to 4,000 (on its left), therefore is rounded off to 5000.

Exercise 23: Complete the table below:

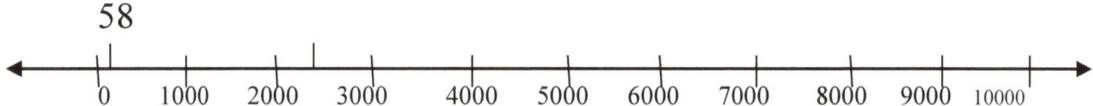

Numbers	Location Between	Round up/Round down (hint: look at the Hundreds line digit)	Answers (rounded-off to the nearest 1000)
58	Down 0 — Up 1000	Round down	0
5, 337	5000 – 6000	Round down	5,000
7, 688	7000 – 8000	Round up	8,000
9,966			
4,200			
58,113			
9,044			

8.5 Further Tips For Rounding Off Any Number:

➢ Circle the digits of the number up to the value that they want you to round off the number to, starting from the digit with the highest value. For example, if asked to round off 5679 to the nearest ten, circle all the digits from highest value to the TENS column (i.e. (567) **9.**)

➢ Look at the first digit on the right that comes directly after the digit in the column that you have been asked to round off to (*in this case the first digit on the right that comes directly after the TENS column is 9*).

➢ If that digit which is outside of the enclosed digits is five (5) or more than five, add one (1) to the enclosed digits and change the digit(s) outside of the enclosed digits to zero(0).

For example (5,67)9 ⟶ 5,680

➢ If that digit which is outside of the enclosed digits is four(4) or less than four, let the enclosed digits remain the same and change each digit(s) outside of the enclosed digits to zero(0). This will thus ensure that the value of the enclosed digits remains the same.

What is 3,252 rounded off to the nearest ten?

For example

> What is 64.7 rounded off to the nearest whole number?

In this case, the whole number is 64, so circle it. The digit that comes after it is 7; this is more than 5, so you add 1 to the whole number (rounded –up) and put a zero to take the position of the tenth column.

+1

64 .7 ⟶ 65.0 or 65

Exercise 24

Can you solve these questions below on your own?

1) Round off 8663 to the nearest 100

2) Round off 3218 to the nearest 10

3) Round off 794.5 to the nearest whole number

4) Round off 33.453 to the nearest 10th

5) Round off 62.432 to the nearest 100th

6) Round off 567,984 to the nearest 10,000

NUMBER CONCEPTS:

SOME *THINGS YOU NEED TO REMEMBER*

- **Odd Number** – This is a number that cannot go into groups of two (2) without a remainder (e.g. 1, 3, 5, 1,347…)

- **Even Number** – This is a number that can go into groups of two (2) without a remainder (e.g. 2, 4, 56, 658, 100…)

- **Ordinal Numbers** – These are numbers that show position Example:

1st (first)	4th (fourth)	7th (seventh)	12th (twelfth)
2nd (second)	5th (fifth)	8th (eighth)	20th (twentieth)
3rd (third)	6th (sixth)	9th (ninth)	21st (twenty-first)

- **Roman Numerals –** Numbers represented by letters. Example:

I = 1	V = 5	X = 10	L = 50	C =100	D = 500
IV= 4	IX = 9	XL = 40	CD = 400	M = 1000	

- **Square Number –** This is the result or product of a number that is multiplied by itself. Example 2 x 2 = 4, therefore 4 is a square number; 5 x 5 = 25, there 25 is a square number.

- **Multiples** – repeated addition of the same number (think multiplication tables). Examples:

2 = 2, 4, 6, 8, 10	**3** = 3, 6, 9, 12, 14	**4** = 4, 8, 12, 16, 20
6 = 6, 12, 18, 24	**7** = 7, 14, 21, 28	**8** = 8, 16, 24, 32, 40

- **Lowest Common Multiple (L.C.M.)** – The lowest multiple that is the same for two (2) or more give numbers. Example: What is the L.C.M of 3 and 4?
  Multiples of: 3 = 3, 6, 9, 12   4= 4, 8, 12

Therefore, 12 is the L.C.M. of 3 and 4 because it is the lowest multiple that is the same in both of them. (The ***prime factorization*** method may be used instead to find the L.C.M.)

- **Factors** – These numbers are multiply to gives a product. Example:

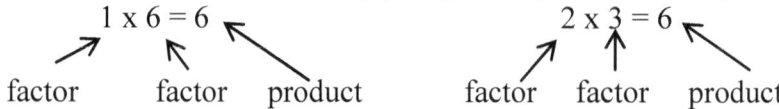

$$1 \times 6 = 6 \qquad\qquad 2 \times 3 = 6$$

factor   factor   product   factor   factor   product

[The factors of 6 are: 1, 2, 3 and 6]

- **Prime Number** – A number that has only two factors; itself and 1. Some examples are 2, 3, 5, 7, 11, 13, 17, 19, 23, 29, 31…

- **Composite Number** – A number that has more than two factors. Some examples are 4,6, 8, 9, 10, 12, 15, 16, 18 …

- **Highest Common Factor (H.C.F.):** This is the largest factor that is the same among a given set of numbers.

- **Place Value:** This tells where the digit is placed. Example:

Thousand	Hundred	Tens	Ones	tenths	Hundredths
1	2	3	4	• 5	6

The **place value** of the underlined digit:
1234.56= thousands   1234.5= hundreds   1234.56 = ones

1234.56= tenths   *(NB. The decimal point separates the wholes from the fractional part.)*

- **Value:** This tells how much the digit is worth. Example

Thousands	Hundreds	Tens	Ones	tenths	Hundredths
1	2	3	4	• 5	6

The **value** of the underlined digit:

1234.56= 1000   1234.5 = 200   1234.56 = 4   1234.56= 5/10 (five tenths)

- **Number Sequence/Pattern**: Find out what is being done from one number to the other next to it; it is addition, subtraction, multiplication or division. If so, by how much?

Example:

2, 4, ___, ___   (2 is being added each time)

12, 8, ___, ___   (4 is being subtracted each time)

___, 10, 15, ___ (5 is being added each time as you count from left to right; but is being subtracted as you count from right to left)

- **Comparing Numbers Signs:**

< less than (think **left** hand)   a. 12 < 21

> more than (think **right** hand)   b. 134 > 124

= equal (the same)   c. 2 x6 = 3 x4

                                            12   12

MASTERY TEST 1

NUMBER CONCEPTS

NAME: _____

CLASS: _____   **DATE**: _____/_____/_____
                                        D   M   YR

A. Fill in the Missing Counting Numbers

a)._____, 58, _____   b). ____, 99, _____

c). ____, 109, ____   d). ____, 1,000, _____

e.). 30, ____, ____   f.). 79, ____, ____

g). _____, _____, 5012   h). ____, ____, 511

i) 1110, _____,_____

B. Write the Odd Numbers between 13 and 26:

{_____}

C. Write the Even Numbers between 792 and 806:

{_____}

D. Spell these Ordinal Number names:

a) 1st _____   b) 22nd _____   c) 73rd _____

d) 9^{th}_____   e) 112th _____   f) 34th _____

g) 20th _____   h) 40th _____   i) 90th _____

E. Complete correctly with the missing Roman Numerals or Hindu-Arabic Numbers

ROMAN NUMERALS	HINDU-ARABIC NUMBERS
I	
IV	
	7
	9
XVIII	
XII	
	14
	5
	190
MMX1	

F. Write 3 Multiples of:

a). 4 _____   b). 6 _____   c) 8 _____

G. What is the Lowest Common Multiple (LCM) of:

  a) 3 and 6?   b) 2, 8 and 4?   c) 2, 7 and 28?

H. Write 5 Prime Numbers:_____

I. Write 5 Composite Numbers:_____

J. List the Factors of:

a)12 _____   b) 28 _____

K. What is the Greatest Common Factor of 12 and 18?_____

L. In the number 4,635.129, what is:

a) the value of the digit 2? _____

b) the place value of the digit 2? _____

c) the digit with the lowest value? _____

d) the digit with the highest value? _____

e) the number in its expanded form?

_____

f) its numeral name?

_____

M. Write the place value of the digit 2 in each number below:

a. 1,327 =   b. 2, 569 =

c. 32 =   d. 7,215 =

N. Circle the number below that has a 5 with a value of fifty.

a. 225   b. 5473   c. 2,358   d. 4,580

O. Complete the number sequence (number pattern):

a) 4, 6, ____, ____   b) 6, 12, ____, ____

c) 3b, 5c, ____, ____   d) ____, 25, 28, ____

P. Write these numbers in words:

a) 101

_____

b) 110

_____

c) 65, 040

_____

d) 828,399

_____

e) 4. 47

_____

Q. Write these number names in figures:

a) fifty four _____

b) six hundred and eighty eight _____

c) two thousand and one _____

 d) thirteen _____

e) seven thousand, one hundred and fourteen _____

f) twenty five thousand _____

R. Round-off these numbers to their nearest 10

a) 4 =   b) 17=   c) 52=

d) 231=   e) 60 =

S. Round-off these numbers to their nearest 100:

a) 79 =   b) 204 =   c) 973 =

d) 752=   e) 519 =

T. Round-off these numbers to their nearest 1,000:

a) 1,050 =   b) 689 =   c) 3,299 =

d) 67 =   e) 9,645 =

U. Compare these numbers below using the signs:

$<$ $>$ $=$

a) 123 ☐ 132   b) 4567 ☐ 3457

c) 3,333 ☐ 999   d) 2,890 ☐ 2809

e) 2 x 4 ☐ 4 x 2   f) 12 ÷ 3 ☐ 18 ÷ 6

g) 15 + 8 ☐ 27 – 5   h) 2/3 ☐ 1/3

i) 3/4 +1/4 ☐ 1

V. Write these numbers in their Expanded Form:

a) 257 = _____

b) 16 = _____

c) 8,920 = _____ _____

d) 4,566 = _____

e) 12,341= _____

f) 2,678= _____

W. Write these numbers in their standard (shortened) form:

a) 2000 + 300 = _____

b) 5000 + 500 + 20 + 2 = _____

c) 600 + 10 + 7 = _____

X. Complete Correctly:

Exponential Form	Factor Form	Product
	2 x 2 x 2 x 2	
9^2		
	24 x 24	
10^4		

END OF TEST

MASTERY TEST 2

NUMBER CONCEPTS

NAME: _____

CLASS: _____   DATE: _____/_____/_____
                                        D   M   YR

A. Fill in the Missing Counting Numbers

a). _____, 58, _____   b). _____, 9999, _____

c). ____, 1009, _____   d). _____, 111000, _____

e.). 3030, _____, _____   f.). 79, _____, _____

g). _____, _____, 5012   h). _____, _____, 511

i) 1110, _____, _____   j). 8001, _____. _____

B. Write the Odd Numbers between 173 and 191:

_____

C. Write the Even Numbers between 792 and 806:

_____

D. Spell these Ordinal Number names:

a) 21st

_____

 b) 12th

_____

c) 563rd

_____

d)109th

_____

e) 705th

_____

f) 244th

_____

g) 20th

_____

 h) 40th

_____

i) 90th

_____

E. Complete correctly with the missing Roman Numerals or Hindu-Arabic Numbers

ROMAN NUMERALS	HINDU-ARABIC NUMBERS
XL	
CIV	
	90
	1139
XVIII	
XII	
	514
	5000

F. Write Three Multiples of:

a). 4 = _____

 b). 6 = _____

c) 8 = _____

G. What is the Lowest Common   Multiple (LCM) of:

a) 12 and 36?   b) 2, 18 and 62?   c) 21 and 84?

H. List the Factors of:

a) 12 _____

b) 28 _____ .

c) What is the H.C.F. of 12 and 28? _____

I.   Circle the Composite Numbers below

11   15   23   63   90   17   32   53

J. In the number 4,629, what is:

a) the value of the digit 2? _____

b) the place value of the digit 2? _____

c) the digit with the lowest value? _____

d) the digit with the highest value? _____

e) the number in its expanded form? _____

f) its numeral name?

_____

K. Write the place value of the digit 2 in each number below:

a. 1,327 =   b). 2, 569 =

c. 32 =   d). 7,215 =

L. Circle the number below that has a 5 with a value of fifty:

a. 225 =   b. 5473   c. 2,358   d. 4,580

M. Complete the number sequence (number pattern):

a) 4, 6, _____, _____   b) 6, 12, _____, _____

c) 3b, 5c, _____, _____   d) _____, 25, 28, _____

N. Write these numbers in words:

a) 101

_____

b) 110

_____

c) 5, 040

_____

d) 28,399

_____

O. Write these number names in figures:

a) fifty four _____

b) six hundred and eighty eight _____

c) two thousand and one _____

d) thirteen _____

e) seven thousand, one hundred and fourteen _____

f) twenty five thousand _____

P. Round-off these numbers to their nearest 10

a) 4 = _____   b) 17=_____   c) 52=_____

d) 231= _____   e) 60 =_____   f) 45.6 =_____

Q. Round-off these numbers to their nearest 100:

a) 79 =_____   b) 204 =_____   c) 973 = _____

d) 752= _____   e) 519 =_____   f) 688.2 = _____

R. Round-off these numbers to their nearest 1,000:

a) 1,050 = _____   b) 689 = _____

c) 3,299 =_____   d) 67 = _____

f) 9,645 = _____   e) 4,003= _____

S. Compare these numbers below using the signs:

$$< , >, =$$

a) 123 ☐ 132   b) 4567 ☐ 3457

c) 3,333 ☐ 999   d) 2,890 ☐ 2809

e) 2 x 4 ☐ 4 x 2   f) 12 ÷ 3 ☐ 18 ÷ 6

g) 15 + 8 ☐ 27 – 5   h) 2/3 ☐ 1/3

T. Write these numbers in their Expanded Form:

a) 257 =   b) 16 =

c) 8,920 =   d) 4,566 =

e) 12,341=   f) 2,678=

U. Write these numbers in their standard (shortened) form:

a) 2000 + 300 = _____

 b) 5000 + 500 + 20 + 2 =   _____

c) 600 + 10 + 7 = _____

END OF TEST

CERTIFICATE OF ACHIEVEMENT
in
Mathematics

Mastering Number Concepts
at Grade _____level

This certificate is presented to

_____

on _____day of _____, 20_____

Signed by   Signed by

_____   _____

INDEX